Kirill Levine

THERMODYNAMICS AND PHYSICAL CHEMISTRY OF SURFACE

Textbook with examples and problems

Science Impact
USA Charleston SC
2018

Thermodynamics and physical chemistry of surface
Textbook with examples and problems

This textbook can serve material for few lectures in thermodynamics and physical chemistry. Recommended for teachers reading lectures of thermodynamics and for students. In details are discussed the First and Second law of thermodynamics, mechanisms of nucleation and growth at the surface.

© All rights reserved. No part of this book can be reproduced, stored in a retrieval system or transmitted in any form or by any means: electronic, electrostatic, magnetic, tape, mechanical photocopying, recording or otherwise without the written permission of the Publisher.

Editor: prof. A.G. Syrkov

Kirill Levine, St. Petersburg Mining University, Faculty of Fundamental and humanitarian sciences, Department of General and Technical Physics

Published by Science Impact, South Carolina, Charleston, USA. Email: impact_press@hotmail.com

Science impact, SC, USA, 2018

© Kirill Levine, 2018

УДК 541.1(075.8)
ББК 24.5

About the author

Author graduated from radiophysics faculty of Leningrad's Polytechnic University. Obtained PhD at the University of Cincinnati (UC) (Ohio, USA) at the department of Materials Science and Chemical Engineering. Worked as post doctoral research associate at North Dakota State University (NDSU) (North Dakota, USA) specializing in electrochemistry and corrosion protection. Later was lecturing Analytical Chemistry at the same university. Later was teaching at St. Petersburg Polytechnic University developing discipline "physical chemistry of nanostructured materials". At present teaches at St. Petersburg Mining University all areas of general physics for students of 1-2 years of study. Since 2010 is Editor in Chief of journal "Smart Nanocomposites" (Nova Science Publishers, NY).

Introduction
Thermodynamics and nanotechnologies

This is well known that thermodynamics is a science describing whether certain processes can occur in principle. It does not say anything regarding the timing of a specific process. The example is aluminum oxidation reaction. Thermodynamics predicts the rate of oxidation of pure aluminum to be very high (gram in less than an hour). Practically, industry uses details from aluminum alloys for decades. This is due to oxide film covering aluminum from oxygen and therefore preventing corrosion.

In this example thermodynamics gives wrong prediction. At the contrary, talking about gas laws, thermodynamics describes the short-term system behavior very precisely.

Why nanotechnology approaches actively utilize thermodynamics? The answer is in the possibility of predictions allowing by thermodynamics. But approaches of thermodynamics in their application to nanoscale have to be essentially corrected comparatively to classical sizes.

Based on the development of micro-electronics, in 1959 basic ideas of nanotechnology were formulated by Richard Feynman in his talk given during the meeting of American Physical society at Caltech. The slogan of his talk was: "there's plenty of room at the bottom" [4].

In Western world ideas of nanotechnology were formulated more than 50 years before by graduate of Mining university P. Weimarn who in 1908 – 1915 basing on studying of colloidal heterogeneous systems formulated that between the world of micro and macro there is something in between which can be called "nano", which size belongs to area 10^{-5} и 10^{-9} m [1]. Therefore, conceptually he formulated basics of nanotechnology [2, 3].

Main laws of thermodynamics

Any surface can be defined as a phase boundary.

*[1] **Phase**. Phase is a part of a system uniform by its physical and chemical characteristics, divided from other parts of a system by boundary surfaces.*

Ideal boundary surface prevents heat and mass transfer across it. An example of such surface is a surface with zero heat conductivity. In real life every surface possesses finite heat conductivity greater than zero, for example thermos or Dewar flask.

Let's call *isolated system a system surrounded by ideal boundary surfaces.*

First law of thermodynamics

According to the first law of thermodynamics, energy in isolated system possess a constant value. (Energy does not appear or disappear, it flows from one form to another).

***First law of thermodynamics** states that amount of heat δQ, transferred to a system, is spent to changing intrinsic energy of a system dU and work δA, made against extrinsic forces.*

$$\delta Q = dU + \delta A \qquad (1)$$

---

[1] By this sign definitions are highlighted.

The First law of thermodynamics declines the possibility of a machine of a perpetual motion (perpetuum mobile (*lat.*)) which makes more work than the quantity of heat transferred to it.

Example:
Engine of internal combustion can be considered as an isolated system. Heat is transferred to it through the burning of a fuel. It is consumed to increase the temperature of an engine (at least when the engine is not hot enough) and work against forces of inertia and friction.

Entropy

Entropy (*S*) is one of the most important quantities of thermodynamics. It is related to so-called coordinate state of matter. An example of a coordinate state of matter is volume of a mechanical system (mechanical coordinate state), chemical coordinate is amount of moles of matter. By analogy, entropy is a heat coordinate of matter. According to thermodynamical definition:

Entropy is a quantity, the entire differential of which is equal to zero in a reversible process.

Reversible process is a process, which initial state is equal to the final.

According to statistical definition,

Entropy is a statistical weight of a system (number of all possible states in which the system can occur).

From statistical definition it follows that entropy is a characteristic of disorder.

Example

Imagine system, composed of three *binary cells, cells, which state can be characterized either as null or unit.* Each cell can be either empty or filled. Let's count the number of all possible states of this system. The elementary calculation shows that it equal to 8.

0	0	0
0	0	1
0	1	1
1	1	1
1	1	0
1	0	0
1	0	1
0	1	0

According to statistical definition, entropy of this system will be equal to

$$S = \ln 2^N = N \ln 2 \qquad (3)$$

Quantity defined this way is dimensionless because logarithm is dimensionless. Even if we would not take logarithm, the number of all possible states of a system would also be a dimensionless number.

This is impossible to directly measure entropy, but it is possible to measure the temperature T and write it for the elementary amount of heat Q:

$$\delta Q = TdS \qquad (4)$$

The first conclusion that can be driven after analyzing this formula is that the temperature must possess dimension of energy, which is Joule. But this is not true. Temperature possesses dimensions of degrees, which in SI system are Kelvins. The solution for this paradox is in historical origination of the quantity of «temperature». The temperature was measured relatively to changes of states of matter: melting and evaporation, which occurred between two fixed points: temperatures of melting and evaporation. This way Reaumur, Fahrenheit and Celsius temperature scales originated. Between fixed points the temperature was determined as a measure of thermal expansion of such matters as alcohol, water, or mercury. Those sorts of temperature were not linked to understanding of intrinsic energy of matter, as they appeared before works of Joule, Thompson and Mayer. Later Kelvin temperature scale was introduced, which had only one fixed point: the so-called temperature of an absolute zero. Due to historical traditions, this scale also possessed dimensions of degrees.

Absolute zero is the temperature, at which any thermal motion of particles (atoms or molecules) stops.

Temperature depends on intrinsic energy as:

$$E = k_b T \qquad (4)$$

where k_b – Boltzmann constant $1{,}38 \cdot 10^{-23}$ Дж/К.

From (4) follows that upon supplying of heat, entropy increases. As was mentioned before, direct measurement of entropy resembles principal experimental difficulties. More convenient to experimenter quantity of heat capacitance C is used:

$$\delta Q = C dT \qquad (6)$$

Specific heat capacitance is an energy, which has to be supplied to a body of a unit mass to increase its temperature by one degree.

Second law of thermodynamics

According to the **Second law of thermodynamics**, heat cannot be transferred from less heated body to more heated.

From the Second law follows the impossibility of the creation of a so-called perpetual motion engine of the second kind, the one which is capable of making work with the help of transferring heat from more cold body to a less cold one.

The Second law can also be formulated as a law of increasing entropy in isolated system:

$$\delta S \geq 0 \qquad (7)$$

or in words: entropy in isolated system cannot decrease.

Theory of heat death of the Universe

Basing on (7) in 1852 Thomson (lord Kelvin) suggested theory of a heat death of the Universe, with which in 1865 agreed Clausius. This theory is popular until now and disputes around it still continue. According to this theory since the time passed from Big Bang (13,72 billion years), cooling of Universe takes place. Indeed, if to agree that the consequence of Big Bang was formation of molecular hydrogen, the following evolution of the Universe can be imagined as condensation of hydrogen to stars, burning of stars, consuming hydrogen, and releasing of energy to outer space. According to this theory this is logical to assume that when stars will burn out all hydrogen, and formed more heavy elements in nuclear reactions will give birth to another even more heavy elements, and this is until the end of periodic system, further heat release will become impossible making impossible further continuation of life.

Findings of modern astronomy, however, do not confirm theory of Thomson and Clausius. In different areas of the Universe, including our galaxy, take place active processes of stars formation from atomic hydrogen. For example, in Milky Way this is an active star creation in Orion nebula [5]. Considering large age of the Universe it can be doubted that atomic hydrogen, which was created practically in the moment of Big Bang (from 10^{-34} to 10^{-32} s from Big Bang), in the process of a so-called baryogenesis[2] [6], until now have not completely condensed to stars long time ago. Therefore it can be assumed that at some corners of Universe (and Orion nebula is one of them) there are still processes of the creation of matter in the form of atomic hydrogen. Physics of the creation of matter, if this process takes place in reality, is not understood yet. However, the fact of the possibility of it doubts the theory of a heat death of the Universe.

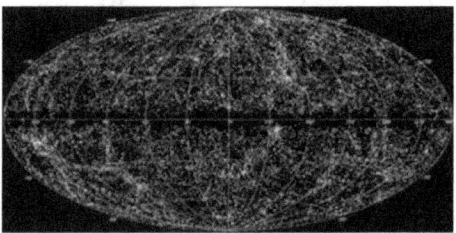

---

[2] Baryogenesis is joining of quarks and gluons in baryons, such as protons and neutrons.

Maxwell daemon

«Maxwell daemon» is the title of imaginary experiment suggested by Maxwell in 1867 to bypass, at least theoretically, the Second law of thermodynamics. Let's imagine gas in a vessel divided in two parts with a small window between parts. There is an intellectual system allowing "fast" molecules to flow from the left-hand to the right-hand side of a vessel (Fig. 1) opening window every time when "fast" molecule comes, and closing it for "slow" molecules. After passing of some time, between the left and right-hand sides of a vessel there settles a pressure difference, which can be utilized for making work. This is a paradoxical situation of bypassing the Second law.

Entropy of gas with separated fast and slow molecules is less than of the initial gas. But from the law of non-decreasing entropy in isolated system, entropy of daemon has increased. Reaching certain value, our intellectual device will no longer be able to make its work due to too high entropy. But in the left-hand side of a vessel there are already less fast molecules than that in a right-hand side. Therefore, some work can be done by restoring the heat balance. Isn't it an another paradox?

Contradictions were resolved by Szilard in 1929. Explanation in simple way can be provided from the modern theory of informatics.

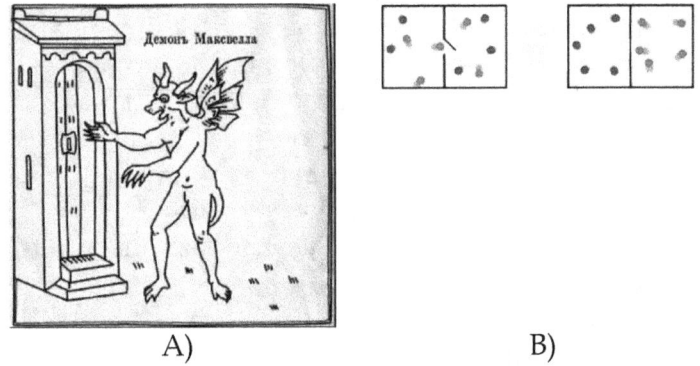

Fig. 1.
A) Illustration from ancient encyclopedia. Maxwell daemon closes the door for entropy.
B) Modern view at work of Maxwell daemon, separating "cold" and "hot" molecules.

It has to be agreed that "daemon" has created certain pressure difference between the left and right-hand parts of a vessel. Entropy of "daemon" has increased. But he did not make any work, he was only making computations to open and close the window at a proper moment. The consequence of this computing was work that can be measured in Joules. If daemon is a computer, than making computations he was scribing information to some carrier. Now, to erase this information, he has to spend energy exactly equal to one that was gained as a result of his calculating activity. Alternatively, daemon can decrease its entropy (take a brake) by violating the limitation of an isolated system.

Third law of thermodynamics

To make the description more complete, the Third law of thermodynamics can be formulated. It is also known as *Nernst theorem: upon cooling entropy decreases approximating to a certain value, which at zero degrees of Kelvin is equal to zero* [7, 8].

Walter Nernst   William Kelvin

Principal relations of thermodynamics
Enthalpy

Substituting (4) to (1) and writing work as

$$dA = pdV \qquad (8)$$

can be obtained equation that is frequently called the main equation of thermodynamics:

$$dU = TdS - pdV \qquad (9)$$

Considering real (not ideal) physical system interaction between its components has to be accounted. Introducing quantity of intrinsic energy (U) it is worth noting that this is the energy of interaction of all kinds of particles, from which this system is composed. Heat energy is one of the components of intrinsic energy. For convenience, quantity of enthalpy (H), and Gibbs free energy (G), are introduced. They are related as:

$$H = U + pV \qquad (10)$$

Enthalpy is convenient for describing energy of an opened system, such as reaction in an opened vessel at atmospheric pressure.

Enthalpy is equal to changing of energy of an opened system upon adding (or removing) mass without considering internal (potential or kinetic) energy of added mass.

By differentiating (10) and substituting dU from (9), can be obtained:

$$dH = TdS + Vdp, \qquad (11)$$

Which is formulation of the First law of thermodynamics using quantity of entropy.

For describing chemical reactions Gibbs free energy is frequently used. It is also called *"free energy"* or *"free enthalpy"*. Negative change of ΔG means that the reaction can occur spontaneously, however, this is not a sufficient condition for this process.

Substituting Vdp, from (11), to (12) can be obtained:

$$dG = -S\,dT + V\,dp, \qquad (12)$$

$$G = H - TS. \qquad (13)$$

G is a so-called characteristic function, same as U, H u F (Helmholtz energy)

$$dF = -S\,dT - p\,dV, \qquad (14)$$

Which by analogy can be written as:

$$F = U - TS \qquad (15)$$

because those functions characterize the condition of a system. Taking derivative of either of characteristic functions Ψ_i with respect to number of moles, quantity of a chemical potential μ_i of a function i can be introduced. Chemical potential possess some analogy with the electrical one. If a potential is equal to zero, current (chemical reaction) does not go through.

Hermann Helmholtz　　　Josiah Gibbs

Surface thermodynamics
Relation to nanotechnologies

Discussing surface thermodynamics, not just surface boundaries have to be considered, but related to them phase diagram. For example, diagram for contains so-called triple point (point at which three states of matter co-exist without converting to each other). In Fig. 2 this point is shown with circle and has coordinates T = 273.16 K, p = 0.062 M Pa. This triple point is also the way how zero degree in Celsius scale (0°C) is determined.

Considering self-organization, exactly in triple point is the possibility of occurring formation of solid nuclei from vapor or liquid. Left upper curvature shows the possibility of formation only from liquid (ice), lover left part shows formation of a solid phase from vapor (snowflakes).

In phase diagram of Fig. 2 is absent, for example, entropy or volume. It does not mean that changing entropy or volume the state of matter cannot be changed.

Similarly to diagram in Fig. 2, phase diagram for any couple of variables (p, T, V, S, n...) can be built. Choice of proper variables has to be made from the considerations of the best description of studying system. For example to describe alloying of germanium (Ge) by arsenic (As), phase diagram in p, T coordinates will be the best option, because As is added to Ge from gaseous phase. But to describe the work of a gas compressor, T, V would be the best option.

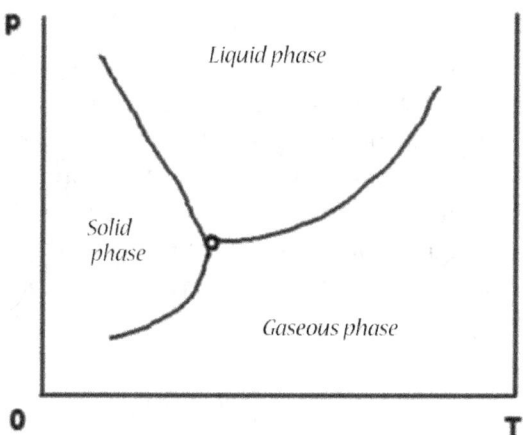

Fig. 2. Phase p, T – diagram of states.

Sufficiency of the describing of a one-componential system by two variables follows from the more general rule of the description of phase transitions of the first order.

Phase transitions of the first order are phase transitions following with changing of a state of matter.

Example of those phase transitions are melting, freezing, evaporation.

Phase transitions of the second order are phase transitions occurring without changes of a state of matter.

Example of second order phase transitions is transferring of ice between amorphous and crystalline forms.

First order phase transitions are following with releasing or absorption of the essential amount of heat. The temperature at which this transition occurs is called *temperature of phase transition*. Also, first order transitions are followed by sharp changes of extensive quantities: volume and entropy, making Gibbs energy exclusively convenient for their description.

$$V = -\left(\frac{dG}{dp}\right)_T \quad (16)$$

$$S = -\left(\frac{dG}{dT}\right)_p \quad (17)$$

Gibbs rule of phases was formulated as a rule, defining number of degrees of freedom for multicomponental system as:

$$p+F = C+2 \qquad (18)$$

where: p – number of phases, F – number of degrees of freedom, C – number of components.

For Gibbs rule of phases this is true that it includes not a voluntarily chosen number of components, but only the necessary minimum required for the description of the system. For example, for the diagram in Fig. 2 the triple point possess zero degrees of freedom in T, P coordinates, because solid, liquid and gaseous phase in this point co-exist. This holds with Gibbs rule (18): Number of components is equal to 1 (H_2O), number of phases – 3,

$$F = 1 - 3 + 2 = 0.$$

Those points at diagram are called invariant points.

Describing surface two factors have to be considered:

1. Large surface are of nanostructured materials.

2. Non-symmetry of local electrical fields at the surface, resulting in the appearance of surface tension forces orienting molecules in such a way that physical or chemical bonding can occur.

Processes at the surface and on surface layers

Surface tension

Surface tension results from non-equity of forces from the bulk and on the surface, tending the body to obtain the smallest possible volume. For a drop of liquid this is a sphere. Surface tension forces result in creating force F in the plane of a surface. To illustrate those forces lets take frame with a wire (Fig. **3**). Surface tension γ, measured in N/m or J/m², is called force, repelling the increase of a surface area.

$$\gamma = \frac{F}{2l} \qquad (19)$$

where l – wire length.

Multiplier 1/2 accounts for the existing of lower and upper surfaces. Behavior of liquid in contact with solid is governed by balance of two phenomena's surface tension and wettability.

Observing behavior of water, moving up in a thin capillary (Fig. 4), a meniscus of radius r can be noticed. This meniscus is caused by wetting forces.

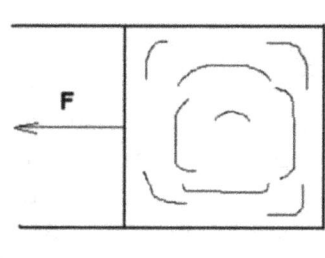
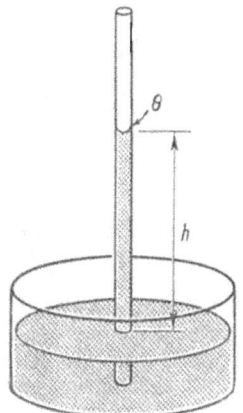

Fig. 3. Scheme showing surface tension in liquid. Force F stretches liquid in a wire frame.

Fig. 4. Lifting of a liquid in capillary.

Boundary between liquid, solid and gas can be characterized by contact angle θ. For pair water/glass θ is equal to zero degrees, for mercury/glass θ is equal to 180°. In the case of a good wettability, surface tension will pull liquid up at heights h until its force is equilibrated by gravity (Fig. 4). If wettability is not very good, $\theta > 0°$, than:

$$2\pi r \gamma \cos\theta = \pi r^2 \rho g, \qquad (20)$$

where g – gravity acceleration,
ρ – density.

Thermodynamics of one-componental systems with surface boundary

To write equations for surface thermodynamics, we will separately account for functions per unit area A (with upper index σ), and per volume (with lower index b). Distortion of a surface will be neglected. Than it is possible to write:

$$H = H_b + Ah^\sigma \tag{21}$$

$$S = S_b + As^\sigma \tag{22}$$

$$G = G_b + Ag^\sigma \tag{23}$$

where:
$$G_b = H_b - TS_b \tag{24}$$

and
$$g^\sigma = h^\sigma - Ts^\sigma \tag{25}$$

For one-componential system surface tension γ is equal to reversible work ω_s, required for increasing surface area by one unit at constant temperature and pressure. Multiplying by l, and considering just one side:

$$\gamma = \left(\frac{d\omega_s}{dA}\right)_{T,p} \tag{26}$$

In isothermal and isobaric processes work is equal to Gibbs energy. Indeed,

$$dU = dq + d\omega \qquad (27)$$

where: dq – infinitely small amount of heat, $d\omega$ – infinitely small work [8].

Taking differential of

$$G = H - TS,$$

substituting here (27) and replacing TS, can be obtained:

$$dG = dU + pdV + Vdp - TdS - SdT \qquad (28)$$

At constant temperature and pressure it becomes:

$$dG = dq + d\omega + pdV - TdS \qquad (29)$$

If changing is made throughout a reversible process, and heat dq is transferred from reservoir kept at the temperature of a system, than

$$dq = TdS \qquad (30)$$

and

$$dG = d\omega_{rev} + pdV \qquad (31)$$

Index "rev" states for reversibility. At constant pressure therefore:

$$dG = d\omega_{rev}. \qquad (32)$$

Coming back to (26), because of (32),

$$\gamma = \left(\frac{dG}{dA}\right)_{T,p} = g^\sigma \qquad (33)$$

Indeed,

$$\left(\frac{dG}{dA}\right)_{T,p} = \frac{\partial(G_b + Ag^\sigma)}{\partial A} = \\ \frac{\partial G_b}{\partial A} + A\frac{\partial g^\sigma}{\partial A} + g^\sigma \frac{\partial A}{\partial A} = g^\sigma \qquad (34)$$

from (28) from

$$dG_{T,p} = dU + pdV - TdS \qquad (35)$$

that is equal to

$$dU - dU = 0.$$

By analogy,

$$\gamma = \left(\frac{\partial \gamma^\sigma}{\partial T}\right)_P = -s^\sigma \qquad (36)$$

After substitution:

$$h^\sigma = \gamma - T\left(\frac{\partial \gamma}{\partial T}\right)_P \qquad (37)$$

This expression contains quantities that can be measured practically, and therefore it is convenient for illustration of thermodynamic quantities. For water at 20°,

$$C\gamma = 0{,}07275 \text{ J/m}^2,$$
$$(\partial \gamma / \partial T)_P = -1{.}78 \cdot 10^{-4} \text{ J/(m}^2 \text{ K)},$$

And surface enthalpy is equal to 0,1162 J/m². Surface enthalpy is an amount of energy related to disappearing of a liquid surface. From everyday experience it is known that at evaporation of 1 м² of liquid surface energy is consumed. At the contrary, upon condensation it is released. Method of determination of surface area of crystalline materials is based on this phenomenon. Crystalline material is kept in vapor, where surface condensation takes place. After that material is immersed in a liquid, and by changing temperature of a liquid, surface area is determined.

Problems for self-control:

1. Prove (36) and (37).
2. Show that

$$\left(\frac{\partial U}{\partial S}\right)_V = \left(\frac{\partial H}{\partial S}\right)_P \text{ и } \left(\frac{\partial H}{\partial P}\right)_S = \left(\frac{\partial G}{\partial P}\right)_T$$

Adsorption and desorption

Formation of condensation centres

Very small energy difference between tending of microscopic drops to evaporate and vapor to condense into nano-sized droplets, defines formation of ordered nuclei's at this surface. Knowing laws of adsorption and desorption allows ruling this process.

Let's describe changes in free energy during vapor condensation. At first we will calculate ΔG behavior during condensation. Proceeding from (12), differentiating it with respect to pressure at constant temperature gives:

$$\left(\frac{\partial G}{\partial p}\right)_T = V \qquad (38)$$

Integrating this equation for solid, where volume does not depend on pressure, can be obtained:

$$G_2 - G_1 = V(P_2 - P_1) \qquad (39)$$

For vapor, considering it as an ideal gas, for n moles can be written:

$$PV = nRT, \qquad (40)$$

$$V = nRT/P, \qquad (41)$$

where R – universal gas constant equal to 8.31 J/(K·mol).

As a result of integration it can be obtained:

$$\Delta G = - nRT\ln(P2/P1), \qquad (42)$$

where minus sign is related to the fact that energy is released. Simultaneously to this, thin film of liquid has a tendency to form droplets of area $4\pi r^2$. Thus from the left-hand side of (33) and (42):

$$\Delta G = 4\pi r 2\gamma - n RT \ln(P2/P1). \qquad (43)$$

n is number of moles in the amount of gas that was consumed to form one droplet:

$$n = 4/3\pi r^3 \frac{N_A \rho}{M} \qquad (44)$$

where: N_A – Avogadro number, ρ – плотность жидкости, M – molecular weight.

Thus,

$$\Delta G = -\left(4\pi r^3 \frac{N_A \rho}{3M}\right) + 4\pi r^2 \gamma \qquad (45)$$

After differentiating (45) with respect to r can be determined critical radius r_c, separating growth and dissolution of droplets:

$$\ln\frac{p}{p^o} = \frac{2M\gamma}{RT\rho r_c} \qquad (46)$$

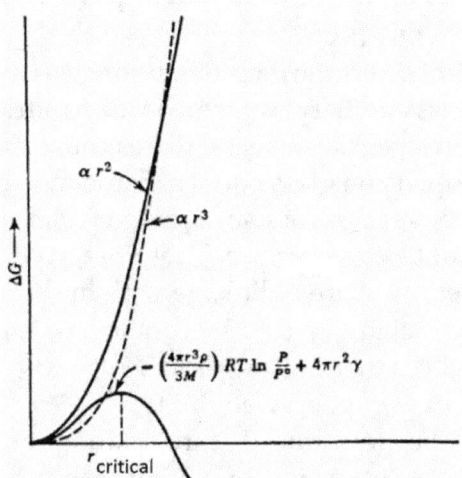

Fig. 5. Changing in isobaric potential during formation of a droplet with radius r from vapor with a constant degree of over-saturation.

Contact angle and linkage to the surface

Meniscus formed by surface of a liquid, for example water in glass capillary, is formed due to surface tension forces. Surface tension forces cause mercury to form droplets on glass surface. Same mercury spreads at copper surface similarly to water spreading onto glass surface. Therefore, wettability can be called measure of linkage with the surface. Wettability is characterized by contact angle θ exceeding 90°, when wetting is absent, and equal to zero in the case of good wettability.

It can be introduced three kinds of surface tensions: surface tension of a liquid on a boundary with gas γ_{lg}, surface tension of a liquid on a boundary with solid, $\gamma_{sl,}$ and surface tension for pair gas-solid γ_{sg}.

At equilibrium there should be equity of interfacial tensions (Fig. 6):

$$\gamma_{lg} \cos\theta + \gamma_{sl} = \gamma_{sg} \qquad (47)$$

or

$$\cos\theta = (\gamma_{sg} - \gamma_{sl})/\gamma_{lg} \qquad (48)$$

It can be noticed that at values of a nominator, when cosine exceeds unit, the equation no longer possess physical sense.

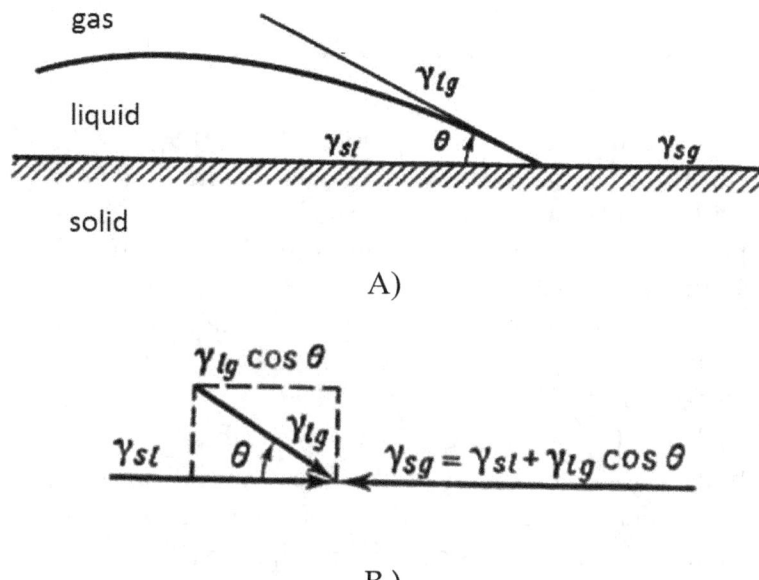

Fig. 6. Liquid in equilibrium with solid surface (A), and vapor (B)

Surface tension in solutions

Surfactants can significantly influence the surface tension of various solvents. Example is fatty acids (soap) and water. Surface tension (isobaric potential of a surface) of water after their addition is significantly reduced.

Hydrophilicity or hydrophobicity of molecules depends on their interaction with the surface (Fig. 7). The same molecule can possess as hydrophilic attracting to water (Fig. 7 A), as hydrophobic, repelling from water (Fig. 7 Б) groups. Molecule, different sides of which possess hydrophilic groups, is called amphiphilic (Fig. 7 C).

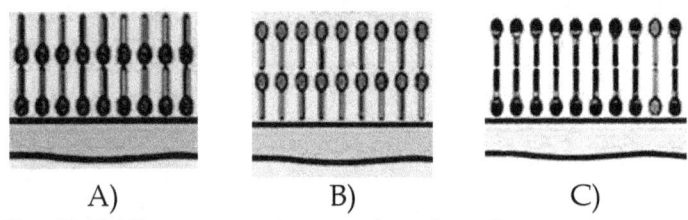

A)   B)   C)

Fig. 7. Different variants of molecules positioning at water surface.

To characterize the influence of a surface active species in solution, Gibbs equation is used.

$$\Gamma_2 = -\frac{1}{RT}\frac{\partial \gamma}{\partial \ln a_2} \qquad (49)$$

where γ - surface and interface tension, a_2 - activity of the component 2. For diluted solutions activity is equal to concentration, therefore:

$$\Gamma_2 = -\frac{c}{RT}\frac{\partial \gamma}{\partial c} \qquad (50)$$

In this representation, surface active matter is considered phase 2, while solution to which it is added is phase 1. Exceeding of the component 2 in solution per unit area is called surface concentration and written as Γ_2. At negative values of the derivative of surface tension with respect to concentration, surface tension is decreased.

Surface pressure

Considering surface of water, covered with surface active species, water surface tension γ_0 will tend to minimize its area to minimum, while as a result of a thin film on water surface will appear force γ.

Than surface pressure π can be written as

$$\pi = \gamma_0 - \gamma \qquad (51)$$

In the case of largely diluted solutions, γ linearly decreases with concentration.

$$\gamma = \gamma_0 - bc \qquad (51)$$

Therefore, in largely diluted solutions

$$\pi = bc \qquad (52)$$

Substituting

$$-\frac{\partial \gamma}{\partial c} = b \qquad (53)$$

And (52) into (49) can be obtained:

$$\pi = \Gamma_2 RT \qquad (55)$$

Thin films at a surface, tending to form monomolecular layer, are called Langmuir-Blodgett (LB) films, named after their discoverers. For those films is true that:

$$\pi = (1/\sigma) RT, \qquad (56)$$

where σ – area, occupied by one molecule.

Following from free spreading of LB films over water surface, Langmuir developed method of estimating film thickness t and molecular surface area a by measuring surface tension with the help of a frame. Surface area occupied by a frame was taken in a moment when surface pressure went through abrupt increase.

Therefore, without using complicated apparatus, area of a single molecule can be determined. Indeed, let's imagine one drop of gasoline spread over area 1 m². Knowing volume of a drop, through Avogadro number amount of molecules I a drop can be calculated. It makes easy to find the cross-sectional area of one molecule keeping in mind than 1. molecule of gasoline at the surface of water is in contact with water by its hydrophilic part, as in Fig. 7 (A), 2. that gasoline layer is monomolecular.

Assignment for self-control

0,106 mg of stearin acid (molecular weight 284, density (0, 85 g/cm³) covers water surface of 500 cm² as a free-standing film. Calculate cross-sectional are of a molecule a.

Adsorption at solids

Adsorption – keeping particles at outer layers of interacting phases, (to be distinguished from *absorption – process, taking place not only at the surface of interacting phases, but penetrating to inner volume of a phase. Absorbing phase* is called *sorbent.* Adsorption takes place because surface energy of particles approaching surface, decreases. Forces causing adsorption are known as Wan-der-Waals forces. Those forces depend on pressure, temperature and concentration. Forces of the same nature cause condensation of a gas to liquid. The degree of adsorption depends on the nature of a solid and adsorbing molecules. Theory of adsorption was developed by Langmuir in 1916. According to this theory, surface of a solid consists of elementary parts, and each of them can adsorb one molecule of gas. Langmuir theory is based on the assumption that:

1. Adsorbing gas in gaseous phase behaves as ideal.
2. Adsorption is limited to monomolecular layer.
3. All elementary parts have equal affinity to gas molecules.
4. The presence of a molecule at one part does not affect properties of neighboring parts.
5. Adsorbing molecules are localized, which means that they don't move along the surface.

At equilibrium, rate of evaporation of adsorbing gas is equal to the rate of condensation:

$$r_\theta = k(1-\theta)P, \qquad (57)$$

where θ - part of a surface, occupied by gas molecules.

r - evaporation rate from completely occupied surface at certain temperature.

Rate of adsorption at the surface is proportional to the area of unoccupied surface (1 - θ) and gas pressure. Therefore, the rate of condensation can be written as k (1 - θ) P, where k - constant at given pressure. This constant contains multiplier accounting that not every molecule in contact with unoccupied surface is adsorbed.

$$\theta = \frac{kP}{r+kp} = \frac{1}{1+k'/P}, \qquad (58)$$

where v_m - volume of adsorbed gas, completely covering the surface, and

$$k` = r/k. \qquad (59)$$

Therefore, volume v is proportional to pressure P at very low pressures, when k`P is much larger than unit.

Deviations from assumptions made by Langmuir result in complicated adsorption dependence (Fig. 8). Upon approaching P/P_0 to unit, volume condensation occurs. At real surfaces also are possible deviations from assumptions made by Langmuir.

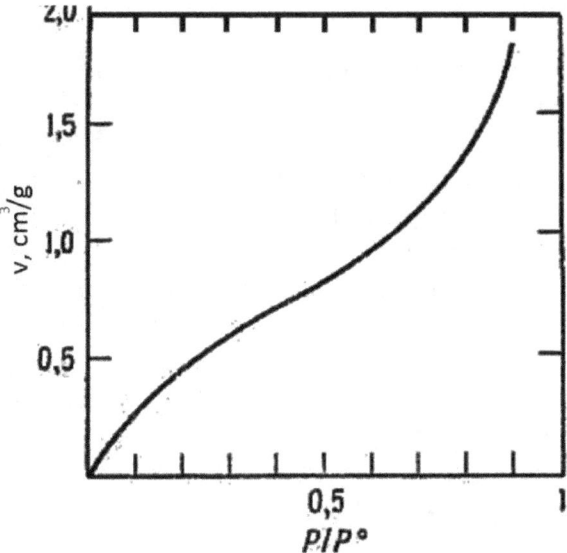

Fig. 8. Nitrogen adsorption at finely grained potassium chloride [9].

In particular, real surfaces can be non-uniform, different faces of a crystal possess different affinity to gas molecules, same as dislocations, vacancies, defects came to surface rate additional adsorption centers. Those deviations can be used in a controlled way to realize surface nanostructures growth.

Assignments for self-control
1. Taking degree of over-saturation P/P^0 for water equal to 4, find critical water droplets radius.
2. Find crystallization radius for chemically pure water, taking its freezing point - 40° C.
3. Explain why surface tension at certain point reaches maximum at Langmuir experiment?

Self-organization of non-organic structures. The role of a surface

Self-organization of ordered nano-sized structures is an example of «smart» behavior of nanostructures. Another samples of such behavior are self-recognition, reproducing, catalysis, self-assembling.

Self-assembling is an evolution of a system in the direction of surface isolation through spontaneous linking of a few (many) components, followed by formation of discrete or composite particles at molecular (covalent), or above-molecular (non-covalent) level.

Self-assembling is characterized by a certain positioning of atoms in a solid.

Self-organization is self-assembling going spontaneously (without applying internal factors, such as electrical fields or temperature gradients.)

Driving force of a self-assembling system is a tendency of atomic system to attain configuration related to lowest potential energy. From those processes on solids the most valuable and important is the process of spontaneous crystallization. Crystalline state of matter is more stable than amorphous. Therefore every amorphous form possess a tendency to crystallize. At the same time it is known that entropy of crystalline phase is less than that of the amorphous. Why the spontaneous process followed by entropy decreasing is anyway possible? The answer becomes clear after the comparison of enthalpy H and TdS of the process. Crystallization is exothermic reaction. Released energy dissipates in surrounding space, therefore, 2nd law of thermodynamics is not applicable here.

Laws of crystallization are defined not only by physical and chemical properties of the media where the process takes place, but also by internal conditions, where media is placed.

$$\Delta G = 4\pi r^2 \sigma^* - 4/3\pi r^3 \Delta g \qquad (60)$$

Changing of free energy occurs non-monotonously with the growth of size (radius) of nuclei (Fig. 9).

Creating nuclei's surface required making work on the system, while formation of crystalline nuclei's volume liberates energy in a system. Changing free energy possess maximum for clusters with critical radius

$$r_{cr} = 2\sigma^* / \Delta g \qquad (61)$$

Formation of clusters with radius less than critical requires positive change of free energy, and system in this situation becomes unstable. In this situation some dynamic equilibrium of the amount of such clusters exists. Clusters with dimensions more than critical possess favorable energy conditions for growth. Rate of crystallites nucleation v_n is proportional to the concentration of nuclei's with critical size and the rate at which those nuclei's are formed:

$$v_n \sim \exp\left(\frac{-\Delta G_{cr}}{k_B T}\right) \exp\left(\frac{-E_a}{k_B T}\right), \qquad (62)$$

Where: ΔG_{cr} – free energy for the formation of critical nuclei, k_B – Boltzmann constant, T – absolute temperature.

Multiplier $\exp(-Ea/k_B T)$ is an impact of atoms diffusion to nucleation followed by further nuclei's growth. It can be characterized by activation energy E_a. As ΔG_{cr} is inversely proportional to T^2, the rate of nucleation changes as $\exp(-1/T^3)$. This is obvious that nucleation of each certain place takes place in narrow temperature interval, below which nothing happens, but reactions shown above take place relatively fast.

Spontaneous crystallization is widely used for creating structures with quantum dots without using lithographic methods. This method is used for the formation of nano-crystallites in non-organic and organic materials.

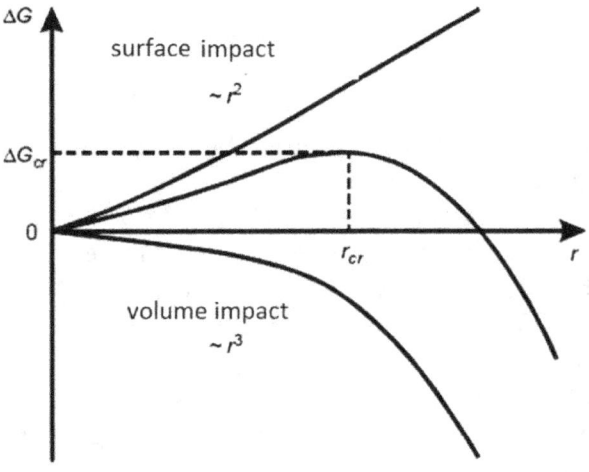

Fig. 9. Changing of free energy of crystalline nuclei depending on its radius.

Conclusion

In this tutorial are reviewed basic principles of thermodynamics. Among discussed topics are first and second laws of entropy, theory of a heat death of the Universe, imaginary experiment known as Maxwell daemon, physical sense of temperature, entropy. Principal thermodynamic relations containing enthalpy and free energy are given.

In surface tension chapter, among discussed topics are surface tension forces, adsorption and desorption on the surface, resulting in formation of nuclei's and nanoclasters.

Questions and assignment for self-control are provided.

Bibliography

1. A. G. Syrkov About priority of Saint Petersburg Mining University in the field of nanotechnology science and nanomaterials // Journal of Mining University, 2016, V.221. P.730-736.
2. P.P. Weimarn, The new classification of aggregate states of matter and the fundamental law of dispersion science // Notes of Mining Institute, 1912 V.4. (2). P.128-143.
3. P.P. Weimarn, I.B. Kagan, A simple common method of obtaining any object in any state of solid colloidal solutions of any dispersity ranging from molecular // Notes of Mining Institute. 1910. V.2. (5). P.398-400.
4. Feynman, R.P. "There's plenty of room at the bottom (data storage)". *Journal of Microelectromechanical Systems.* **1** (1): 60–66.
5. Alain Baudry, Nathalie Brouillet, Didier Despois, Star formation and chemical complexity in the Orion nebula: A new view with the IRAM and ALMA interferometers, C. R. Physique 17 (2016) 976–984
6. V. Mukhanov, Physical foundations of cosmology. Cambridge university press 2005.
7. F. Daniels, F. Alberty, Physical chemistry. Wiley, 1955, 671 p.
8. H.C. Van Ness, Understanding Thermodynamics. Dover ed edition.
9. Keenan A.D., Holmes J.N., J. Phys. Colloid Chem., 53, 1309, 1949.

Contents

Introduction..**4**

    Thermodynamics and nanotechnologies...4

Main laws of thermodynamics...................**6**

    First law of thermodynamics.....................6

    Entropy...7

    Second law of thermodynamics................10

    Theory of heat death of the Universe......11

    Maxwell daemon..13

    Third law of thermodynamics..................15

Principal relations of thermodynamics....**15**

    Enthalpy..15

Surface thermodynamics..........................**18**

    Relation to nanotechnologies...................18

**Processes at the surface and on
    surface layers**............................**22**

    Surface tension..22

**Thermodynamics of one-componental
    systems with surface boundary**........**24**

Adsorption and desorption......................**29**

    Formation of condensation centers.........29

    Contact angle and linkage
        to the surface...32

    Surface tension in solutions.....................34

Surface pressure……..……………..35
Adsorption at solids…………..……...38
Self-organization of non-organic structures………………………….....…41
Conclusion…………………………....…..44

Bibliography……………………….....…46

Author would appreciate feedback related to the contents and design of this book sent to email *levinkl@gmail.com.*

For notes

www.ingramcontent.com/pod-product-compliance
Lightning Source LLC
Chambersburg PA
CBHW030055230526
45471CB00003B/1115